I0223445

Atomic Structure

TEACHING CHEMISTRY IN A DIVERSIFIED CLASSROOM

BOOK 4

By Sammie Jacobs

Copyright

All rights reserved under International and Pan-American Copyright Conventions. By payment of the required fees, you have been granted the non-exclusive, non-transferable right to use the materials provided for the preparation of lessons and during the direct instruction of students.

ATOMIC STRUCTURE. Copyright 2020 ©

Student worksheets and assessments may be duplicated for classroom use, the number not to exceed the number of students in each class. Notice of copyright must appear on all copies as provided on the page.

Activity kits duplicated and assembled as directed should display the copyright markings as prescribed in the directions and on the materials.

No other part of this text may be reproduced, up-loaded, displayed, shared, transmitted, down-loaded, decompiled, reverse engineered, or stored in or introduced into any information storage and retrieval system, in any form or by any means, whether electronic or mechanical, now known or hereinafter invented, without the express written permission of Lavish Publishing, LLC.

First Edition
2020 Lavish Publishing, LLC
Teaching Chemistry in a Diversified Classroom book 4
All Rights Reserved
Published in the United States by Lavish Publishing, LLC, Midland, Texas
Select Illustrations by Sanghamitra Dasgupta
Cover Design by: Victor R. Sosa
Cover Images: CanStock
Paperback edition
ISBN: 978-1-64900-003-3
www.LavishPublishing.com

FOREWORD

Teachers, parents, and educators of all kinds – welcome to my classroom!

Here within these pages, and in fact this entire series, are lessons and activities that I have created and used with my own students over the years presented in an easy-to-use format. Through trial and error, hardship and success, I have learned how to present the difficult world of introductory chemistry to students in scalable terms, ways that lend themselves to a wide variety of learners.

So, if you teach in a small class, large class, average class, have ESL or SPED, or even if you are homeschooling, you will find the flexibility of my lesson plans tailor made for you. Each unit includes a complete calendar plan, lessons for each classroom day for forty-five minutes of instruction each, and even formative and summative assessments to check for learning. Simply use the lessons, scale them as need be, and augment whenever you like.

Be sure to join my Teaching Chemistry PLC on Facebook for even more ideas and sharing, and of course, I hope for the best with you and your students!

Sammie Jacobs

TABLE OF CONTENTS

Unit Tools . . . . Pages 1 – 5

Calendar Key and Usage
Guide Calendar
Blank Calendar
Atomic Structure Vocabulary

Lesson Plans . . . . Pages 6 – 49

Daltons Postulates
Atomic Model Foldable
Periodic Table Parts
Nuclear Symbols
Reading Day
Vegium Lab
Calculating Average Mass
Frootloop Isotopes
Atomic Structure Review

Assessments . . . . .Pages 50 – 64

Daltons Postulates Quiz
Atomic Theories Quiz
Periodic Table Boxes Quiz
Calculating Average Mass Quiz
Atomic Model Quiz
Nuclear Symbols Quiz
Atomic Structure Vocabulary Quiz
Atomic Structure Unit Exam
Answer Keys

Unit Calendar

The **Unit Calendar** is an important organizational tool.

It features a list of vocabulary terms (top section) for the unit, the plan of days (center section), and the student learning goals (bottom section).

A **completed teacher copy** of the calendar with the unit overview is provided. On this version, days are not written as dates, but as lesson days. However, on my copy that I use in my classroom, these have been laid to the school or district calendar and those day numbers are replaced with actual meeting calendar dates. There are five in a row, which is a work week, and days that we do not meet or have instruction time are crossed off.

You will notice that the unit calendars are always a whole number of weeks long and every day has a lesson planned, but often you will have holidays or non-class days. When you have less days, combine lessons where appropriate, or omit lessons that are not included on the student goals and exam if necessary. We also only meet for forty-five minutes per day, so the lessons are designed to fit within that time frame. If you have longer class periods, you may decide to augment the days with additional activities, such as vocabulary games.

The **blank student copy** can be used to customize the calendar to match your available days. It can also be copied and distributed to students, along with the vocabulary lists with definitions.

Calendar ideas:

1. Have students fill in the calendar daily with daily topic as they begin class as a way of making them aware of what the day's lesson will entail.
2. Have students fill out the calendar at the start of a new unit so they have the plan ahead of time and can be proactive with their learning.
3. Give students a completed copy with the appropriate dates and topics at the start of a new unit to save time.
4. Have students highlight each day and the corresponding terms on the vocab list with a different color (color coordinating them) for reviewing later.
5. Have students use the calendar to record assignments for each day and highlight them as they are completed and submitted.
6. Have the students write a question for each day to reflect upon later.

Vocabulary

anion	ions	Rutherford's Atomic Model
atomic number	isotopes	Schrodinger / quantum
average atomic mass	Lewis Dot Diagram	mechanic model
Bohr's Atomic Model	mass number	
cation	neutron	subatomic particles
charge	nuclear symbol	Thomson's Experiments
Dalton's Postulates	octet rule	valence
electron	proton	

Daily Agenda

M 1	T 2	W 3	Th 4	F 5
Dalton's Postulates	Atomic Model Foldable Day 1	Periodic Table Parts	Nuclear Symbols	Reading Day

M 6	T 7	W 8	Th 9	F 10
Vegium Lab	Calculating Average Mass	Frootloop Isotopes	Flex Day and Atomic Structure Review	Atomic Structure Exam

Learning Goals

Target Concept	none	weak	solid
Can name Dalton's Postulates and determine 'correctness' of each			
Can explain Thomson's experiment and contributions			
Can explain Rutherford's experiment and contributions			
Can explain Bohr's nuclear atom and contributions			
Can explain Schrodinger's experiment and quantum mechanic model			
Can explain the difference between isotopes of the same atom			
Can use nuclear symbols and calculate average atomic mass			

© Lavish Publishing, LLC

Vocabulary

anion	ions	Rutherford's Atomic Model
atomic number	isotopes	Schrodinger / quantum
average atomic mass	Lewis Dot Diagram	mechanic model
Bohr's Atomic Model	mass number	
cation	neutron	subatomic particles
charge	nuclear symbol	Thomson's Experiments
Dalton's Postulates	octet rule	valence
electron	proton	

Daily Agenda

M	T	W	Th	F

M	T	W	Th	F

Learning Goals

Target Concept	none	weak	solid
Can name Dalton's Postulates and determine 'correctness' of each			
Can explain Thomson's experiment and contributions			
Can explain Rutherford's experiment and contributions			
Can explain Bohr's nuclear atom and contributions			
Can explain Schrodinger's experiment and quantum mechanic model			
Can explain the difference between isotopes of the same atom			
Can use nuclear symbols and calculate average atomic mass			

© Lavish Publishing, LLC

Atomic Structure Vocabulary

word	definition
anion	____ are negatively charged ions - atoms that take electrons.
atomic number	____ is equal to the number of protons in an atom; used to arrange the period table.
average atomic mass	By taking the weighted percentage of all naturally occurring isotopes of an element, we can calculate the ____, which is the mass found on the periodic table.
Bohr's atomic model	____ is a planetary model in which the negatively charged electrons orbit a small, positively-charged nucleus similar to the planets orbiting the Sun (except that the orb are not planar)
cation	____ are positively charged ions - atoms that give away electrons.
charge	Atoms that give or take electrons and become ions are said to have a positive or negative ____, which causes them to attract or repel other particles that have a ____ similar to a magnet (same word in both).
Dalton's Postulates	In the early 1800's, ____ were published, which gave us a strong definition for nuclear theory, much of which is upheld today.
electron	____ are subatomic particles with a negative charge, no mass value, and are located outside the nucleus of the atom; they are responsible for chemical reactions.
ions	____ are the result of gaining or losing electrons by atoms.
isotopes	Atoms have different numbers of neutrons in their nuclei, which are said to be ____ because their masses will vary.
Lewis Dot Diagram	A ____ is used to depict how many valence electrons surround a given atom - a nuclear symbol surrounded by the proper number of dots.
mass number	____ is equal to protons plus neutrons; it is the average atomic mass rounded t a whole number.
neutron	A ____ is a subatomic particle with no charge, mass is about 1, and is located i the nucleus of an atom.
nuclear symbol	The letter or letters that represent an element are called the ____, which can include details about the structure or parts of that atom.
octet rule	The ____ tells us that all atoms want to have 8 electrons in their valence she Remember that all atoms can only hold 2 electrons in the very first energy leve which is the only completely stable exception to this rule, such as Hydrogen o Helium.

proton	A subatomic particle with a positive charge is called a _____, which has mass of about 1, and is located in the nucleus of the atom.
Rutherford's Atomic Model	_____ of the atom describes it as a small, dense nucleus surrounded by orbital electrons. He used the 'gold foil' experiment to make his discoveries, which were the basis for other scientists who completed the model to represent atom in the way they are understood today.
Schrodinger & quantum mechanic model	_____ attempts to explain the structure of an atom and electron configuration by using the laws of probability to predict the location of electrons. He is famous for his 'cat in a box' experiment.
subatomic particles	The parts that make up an atom are called _____ - known as protons, neutrons and electrons.
Thomson's Experiments	In _____, he used a cathode ray to determine charges and locations of subatomic particles. His theory (called the 'plum pudding model) was later disproved but did serve as a strong catalyst future investigation.
valence	The outer energy level of an atom is called the _____ shell and the electrons that are in it are called _____ electrons. (same word for both).

©Lavish Publishing, LLC

Key to Lesson Plans

Topic: coordinated to the unit calendar - what we are learning about today?
Day: coordinated to the calendar - sequence of presentation out of total available
Unit: coordinated to the unit calendar - unit the lesson falls under

Learning Target:
A postable **learning objective** that gives an outline of what the day's lesson will cover and **artifact** the students will produce for evaluation. These can be used as is or modified to suit your student output or to focus on a different aspect of the lesson, as these often only cover one part of a layered lesson and group of activities.

Student Goals:
At the bottom of the calendar, the students have a list of **goals** they want to meet by the end of the unit, which are covered on the **unit exam**. This section coordinates which goals are being addressed during this lesson. This section also lists the **vocabulary terms** from the calendar that will be defined or needed during the lesson so they can be connected, pre-taught, or directed to study afterwards.

Agenda:
A postable list of what activities this day will include. They are divided into two main types of instruction:

Lecture **(L)**, where the teacher is providing direct instruction and the students are actively listening, taking notes and providing feedback at given intervals.

Activities **(A)**, where the student is producing work of some kind and the teacher is observing and providing support when needed.

Student Materials:
A list of materials that each student, pair, or group, will need to complete the lesson. They will need to be prepared ahead of time and be ready to **pick up** as students enter the room or to be **distributed** at the appropriate time during the lesson.

Generally, if it is a **one-to-one item**, I have them ready and students pick them up as they come in to save time. **Paired and Group materials** can be handed out or placed in a location to have a student go and get while the teacher is preparing some other part or completing a task during transition.

Props:
These are items you generally only need one of and can be hidden to pull out at the appropriate time or placed on a table that the students can visit before and after the lesson.

6

They are support items that deepen the understanding of the lecture and either generate questions or provide answers.

Having props is vital for a wide range of learners, so do not feel limited to what I have and use – explore and add anything that you want to aid in this process, as all students benefit. Save your props as you build them, as they often are used in future lessons to bring concepts forward and tie ideas and understanding together.

I have a set of props that stay on my front table throughout the year, as I pick them up and refer to them often: a plain bottle of water that is labeled H_2O, a bottle that is half cooking oil and half water, a bottle that is water with about a gram of dirt in it, and a bottle that is water with about a gram of corn starch in it. Other props are added and removed as needed.

Actions and Rationale:
These are the key points of the lesson and the reasoning behind what is being said or done. All activities will have a separate directions page if needed, as well as a reproduceable student copy and directions for building all props and student materials.

During each unit, we use a variety of learning and presentation styles that include having the students listen, speak, read and write. There are also a variety of study and practice skills woven into the lessons that students can learn to use in and outside of their Chemistry class.

Topic: Dalton's Postulates
Day: 1
Unit: Atomic Structure

Learning Target:
I can name Dalton's Postulates and determine which were accurate versus inaccurate.

Student Goals:
Students will gain the basis of atomic theory by learning about Democritus, with his naming the atom, and Dalton, who formulated the first concrete theories for atoms, with a side focus on the difference between hypothesis, theory, and how science changes as it grows.

Agenda:
A – Warmup: Calendar
L – Dalton's Postulates
A – Reflection Writing: Hypothesis Vs. Theory

Student Materials:
Journals for Notes
Calendars and Vocab for gluing
Slips of paper or notecards for reflection writing

Props:
Display Calendar if students are copying
Visuals for Lecture – power point, etc.

Actions and Rationale:
Warmup – calendar and new unit. I print the blank calendars and display our unit calendar. The students copy it down while I take attendance. I also have a short article in our text book that I have them read if they finish before I am ready, which is over Democritus and his 'discovery' of the atom, which is fun to tie in with this lesson. When we are ready, I go over the calendar to point out due dates, days for special dress for labs, etc. Then I transition into the lecture by going over the reading to make sure they know where the word 'atom' was created, then transition into Dalton.

Lecture – the beginning of atomic theory. Once we have started off with Democritus, we move to Dalton's Postulates, which I display up on the board. The version we have in our text is quite wordy, so we go through and I mark or highlight the main parts (demo for the students on how to pull out short notes) and we write the adapted version in their journal.

You can also find his postulates online with commentary, so have a look around and choose the one you want to use for this lesson. There are even a few videos if you would rather go that route. Either way, students need these basic portions, which is what the quiz here and the unit exam are based on:

1. *All matter consists of indivisible particles called atoms.* **(This was later found to be false. I like to point out that science is worked on by many scientists, so we are always learning and growing over time.)**
2. *Atoms of the same element are similar in shape and mass, but differ from the atoms of other elements.* **(this is partially false, since isotopes of an element have different masses)**
3. *Atoms cannot be created or destroyed.* **(true)**
4. *Atoms of different elements may combine with each other in a fixed, simple, whole number ratios to form compound atoms.* **(true)**

Activity – hypothesis vs. theory. If time permits, I like to have this little discussion with my students and then they get a chance to write a small reflection. I do a quick google search up on the screen and we read the definition together.

According to Merriam-Webster's online dictionary… "The Difference Between a Hypothesis and a Theory. ... In scientific reasoning, a **hypothesis** is an assumption made **before any research** has been completed **for the sake of testing**. A **theory** on the other hand is a principle set to explain phenomena **already supported by data**."

We talk for a few minutes about this definition versus how we use the word 'theory' in everyday life.

Is it the same? Or when we say theory, do we mean hypothesis? (typically, we mean hypothesis because we lack data to support our claim, so we are really guessing)

Can a hypothesis be wrong? (Sure, they are meant to be tested, not 'proven' so we are using them to guide our research. I like to point out that scientist who sets out to 'prove' his idea correct is off to a bad start, and likely to misrepresent his data or ignore things that don't support his hypothesis, which is not what science is about. Science is about discovering the truth, whether we were right or wrong, and using that knowledge to take the next step, make the next discovery, and make the next hypothesis to be tested over and over again until our evidence is strong enough to support a theory.)

Can a theory be wrong? (Theories stand the test of time because they are based on strong experimental data. However, when new technology is discovered, we MAY find that a theory that was once proven by data based on what we had available then must be modified. That is the beautiful thing about science – it is a living changing thing and there is still much to be learned as our ability and technology changes and grows. However, it is safer to assume that scientific theories are correct because they RARELY become altered or nullified once they have enough evidence to be called a theory.)

For the reflection, I have the students contemplate and write about why understanding the difference in the common use and the scientific 'back up' that goes with the word theory matters. Three to five sentences are plenty, and this can be their exit ticket as they leave out from class, which I collect and read/grade for practice points.

Learning Target:
I can build a foldable to display my learning about the atomic theories and models of Thomson, Rutherford, Bohr and Schrodinger.

Student Goals:
Students must know basic information about these four scientists and their contributing atomic models, including where they lived, how they made their discoveries (experiments), a memory trick, and what their model looked like.

Agenda:
A – Warmup: Dalton's Postulates Quiz
L/A – Atomic Model Foldable

Student Materials:
Legal size paper (8.5 x 14)
Color pencils
Source document or internet access

Props:
Completed foldable or demo version

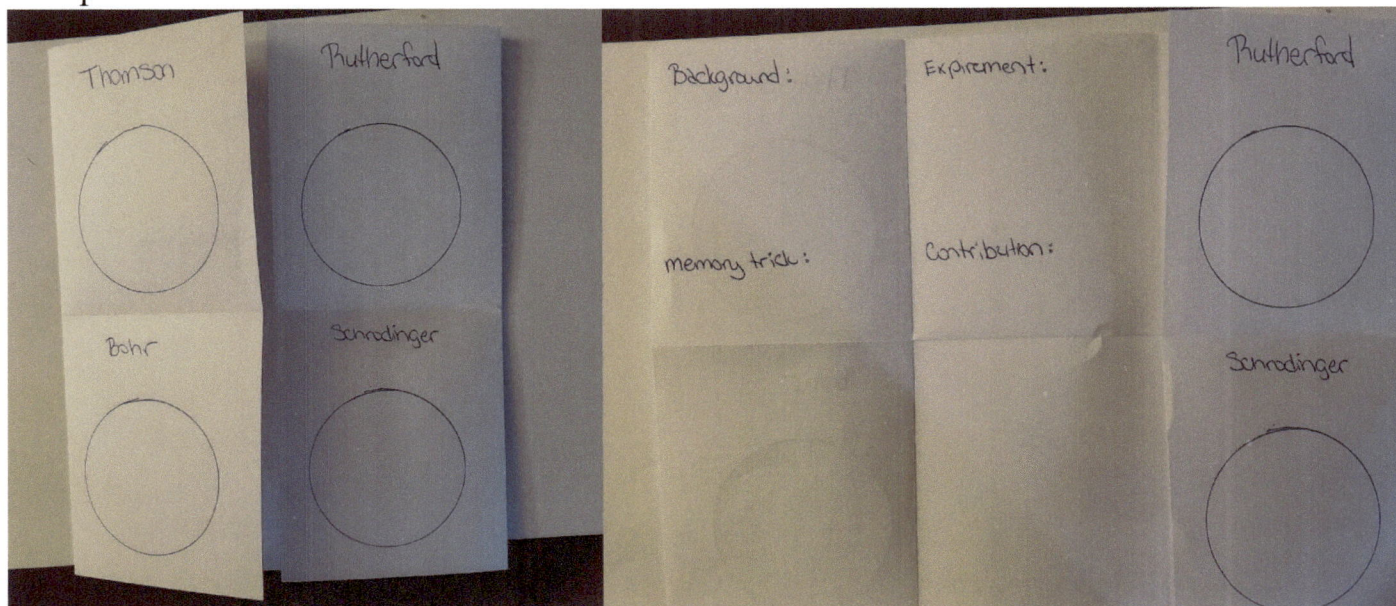

Front	Inside

Actions and Rationale:

Warmup – Dalton's Postulates quiz. They do the quiz while I take attendance, then I take them up for practice points before I go over them. After that, I can transition into the lesson today, which is basically building our main tool for atomic models. The students will turn it in or show it for points, they will use it to study for the exam, and we will eventually glue it in our journal for reference if we ever need it.

Activity – atomic model foldable. For the foldable, I have found over the years to NOT present a completed foldable. If I do, many students simply take it and copy exactly what I had, which defeats the purpose of having them create the foldable.

What I want to see is their thinking, interpreting, and creating this tool they will use to study and prepare for the exam. I have pictures above of exactly what I post up on the board for them to look at.

As for the folding and setting it up, I go through that step by step to help them get a good fold. I call this a 'shutter fold' because the front opens like a set of window shutters to reveal the inside, which we will use a couple of more times this year. In the past, we cut the boxes so they could open one individual scientist / model to see inside, but we found that they became ratty inside their journal, being easily damaged with so many flaps. Therefore, we leave them as whole halves.

I start off the foldable details by talking about the scientific process as a whole. A hundred years before, we had Dalton and his thoughts, but then things got relatively quiet until we had a sudden flurry of discovery, largely due to our advances in technology that made them possible. Although there were others working on these same exact things, these are the four we are going to be focusing on.

On the **front**, they will put the name of the model (scientist) and draw a picture. I always pull out a selection of small cups for them to trace and get nice circles there. I tell them to use color pencils for the coloring, as it will not bleed through and make the inside harder to see.

On the **inside**, they want to pack it with information, but not simply copy their source. Just as we did with the Dalton Postulates, they want to pull out the important parts and put them under the appropriate title: Background, Memory Trick, Experiment, and Contribution.

For your use – this is what they should gather for each of them:

I usually give them this full day to set it up and fill it in. We will use pieces of other days when time allows to work on this. I also like to present the following parts of it over the next few days so it doesn't all happen at once, and I show pictures of their experiments as I talk about them and explain, hopefully after they have already read and formulated an idea of what they are about.

I like to plan ahead which days I am going to talk about each of them– as long as you cover them, it really doesn't matter when or what order so feel free to change them up if need be. Just don't do it today. Let them explore today and build their foldable from what they are able to find by doing their own research. They can add to it later when you give them your 'lecture' for each of these guys.

Atomic Model "Mini-Lectures"

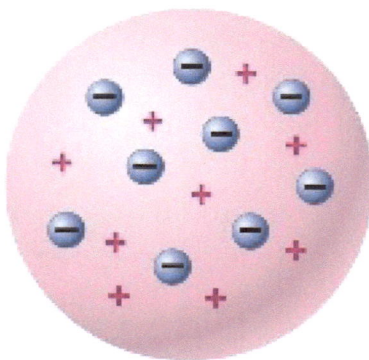

Thomson – cathode ray experiment. I like to display a picture of the cathode ray or have a live one (we have an old one that doesn't actually work, but it is a good visual) and hold up a magnet to show them how moving it over or under would look. We talk about how opposite charges attract and same charges repel, which is going to be a huge concept for us this year. Take your time and make sure everyone understands so they have at least that foundation.

His contribution was electrons with their negative charge, and Thomson's 'memory trick' is historically called the plum pudding model. However, I let my students make up their own as well, because they need to know that name, but they also need one that is actually going to work for them. As I tell my students, memory tricks need to be personal, so having them start learning how to do that for themselves will be a good skill to have throughout the year. Often, we settle on a 'chocolate chip cookie' for this one, but some students like 'pizza' or something else for theirs.

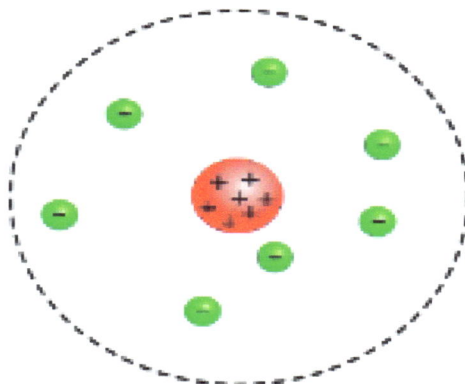

Rutherford – gold foil experiment. I have to use a picture of this one, but it's fun to talk about his shooting the alpha particles at the gold foil. I usually get them to think about aluminum foil (you can even hold up a piece if you want), which helps them picture what that might look or feel like. The picture I have shows the closeup of atoms as large spheres with dense nuclei. The atoms are mostly open space, which allows the particles to mostly pass through, with only those hitting the nucleus of an atom being deflected in another direction. If you can't show a picture, you can always draw one and place arrows to illustrate that action.

His contribution was the dense nucleus with atoms being mostly empty space. The memory trick I suggest for them is the 'tootsie pop' with the tootsie roll center as the nucleus and the candy coating as the electron shells (or cloud) around it, which we will use again later. Students may also come up with their own if they have a better one.

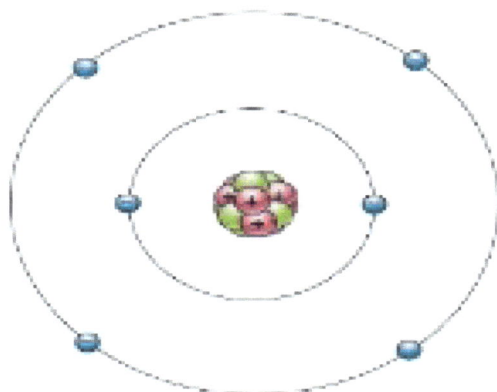

Bohr – his is a model, not an experiment. However, he does a pretty cool trick with his Nobel Prize and a tub of acid, so some of my students like to write about that as his experiment. Mainly, his is the contribution of the Bohr model, which we use over and over to draw what atoms look like in a simple way. I tell my students that I use them as a modified version, and I'll explain more about that when we are ready to draw and use them. For now, students just need to know that the protons and neutrons are packed into that dense nucleus in the center and the electrons are running around it on orbitals (which I liken to tiny 'superhighways'). The memory trick we often use is the solar system, but some like to call it 'orbits gum' or something similar.

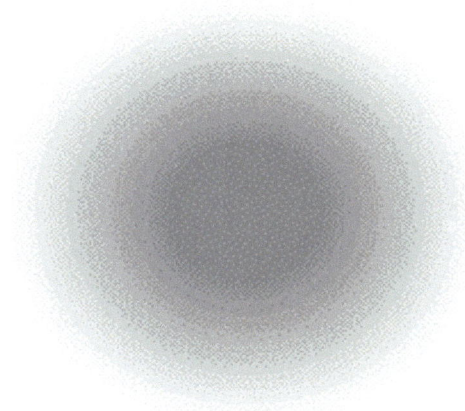

Schrodinger – cat in a box experiment. This is also commonly called the 'quantum mechanic model' so be sure your students pick up on that part. I really enjoy talking about this one, so I start by telling the students that 'no cats were harmed in the performance of this experiment.' It was all theoretical.

Then I tell the students, "Let's work our way around to this. Let's start with dinner. What's for dinner tonight at your house?" You will have many who don't know, but a few may have a solid reply. Then we go through the list of possibilities. Name some foods and restaurants that might be on the menu. Then I ask them, "So, how do you know what's for dinner? When can you know that?" Ultimately, all options are open, so it could be anything – as if all the options were happening at the same time. When the choice is made, that's when you know, whether that means when you get home and see what mom cooked, or you pull into the parking lot at a restaurant, or you place your order with the waitress. That is when you know.

Then I talk about Schrodinger and his cat. Basically, his experiment was to say, "If I have a cat in a box and it is all sealed up, is the cat alive or is the cat dead?" The answer is, the cat is both. Again, both options are happening at the same time continuously, until the choice is ultimately made. Ask, "How do you find out if the cat is alive or dead?" Usually, at least a few will know – you look.

That is how quantum mechanics works – all options available and happening, random and constant, until we peek and see where our electron is. Otherwise, we have zones of 'probable' location. That is his contribution, and it leads to his memory trick – when you have all those electrons buzzing around so fast, it makes a fuzzy cloud looking sphere. Some call it the 'moon' or a 'vanilla wafer' for their memory trick, and that works for me.

Topic: Periodic Table Parts
Day: 3
Unit: Atomic Structure

Learning Target:
I can practice determining element parts, including protons, neutrons, and electrons, using the periodic table.

Student Goals:
Students will need this foundational knowledge for using the periodic table to find key information about elements in general or on average.

Agenda:
A – Warmup: Atomic Model Quiz
L – PT Boxes
L – Atomic Model Foldable: Thomson
A – Magic Board Practice

Student Materials:
Journals for notes
Foldables for Reference / Corrections
Slips of paper for quiz
Reference Periodic Tables
Magic Boards and dry erase makers / erasers

Props:
Either power point or white board to draw the example (I put it up as we go)

Actions and Rationale:
Warmup – atomic model quiz. I like to give this now so the students have an idea of what they should be getting out of the models and what they need to know. They do the quiz while I take attendance, then I take them up for practice points before I go over them. After that, I can transition into the lesson today, which is basically two parts, so do them in whichever order you want – both parts together will be twenty to twenty-five minutes of lecture and that leaves a few minutes for magic board practice at the end.

Lecture – periodic table boxes. This is something students should already know, but often they need a little memory kick to bring it to the surface. It also gives you a chance to bring anyone who never learned it up to speed.

17

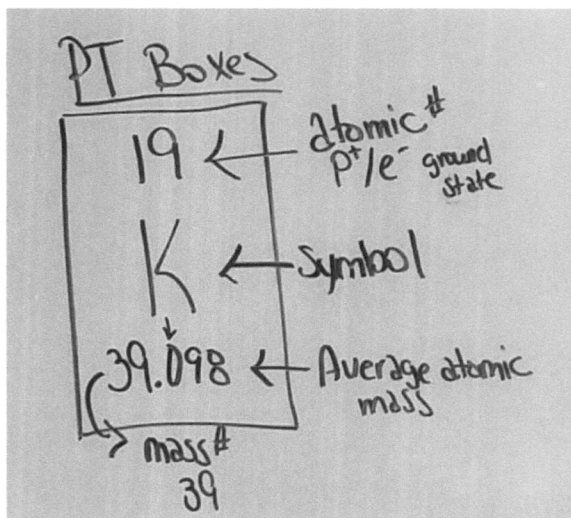

Periodic Table Boxes

1. The boxes represent an 'average atom' and not a specific sample, so some values are not 'perfect'

2. The **Atomic Number** is 'perfect' - it tells the number of Protons and electrons when **neutral** (not ionized) and those values never change

3. **Average Atomic Mass** is not 'perfect' - the values change by sample, so round to a whole number to get the Mass # (these give us isotopes with varied neutrons)

This is a picture of my board after we finished this lecture and next to it is a list of the points that the students need to grasp. Go through each one, but have the students pull out their reference periodic table and practice a few as you go. I always do the 'everyone put your finger on…' routine and walk around watching for participation as we do. They aren't going to learn it if they don't get involved. Do two or three for atomic number – Find Oxygen. What's the Atomic number? Then explain average atomic mass and repeat – find, what is… etc.

Once they can locate things, do a few of rounding that average mass to get mass #. I was surprised how many students struggle with something as simple as rounding, so don't be afraid to draw a little chart on the board for them – 0 to 4 is a 'no bump' and 5 to 9 gets a 'bump' to the next number.

Once you are happy with what they know, you can transition to the next half, which is to talk about Thomson. I have that material for you with the atomic model foldable day, and if you decide you don't want to do his first you can do them in any order, but I like how electrons ties in with the boxes.

Once you have talked about Thomson, you can tie these two things together by pointing out that those electrons are also the atomic number on the periodic table – nice that he discovered them for us. Be sure that you talk about atoms being **neutral.** I like to use the term **ground state** here just to introduce it, which means all the electrons are present (no giving or taking going on), so the number of protons and electrons are equal. No charge exists – it is 0 and you can put those number on the board as + and – to demonstrate that. For example, the atomic number for Oxygen is 8, with 8+ for protons and 8- for electrons. If you add those together, you get 0. That's neutral.

Activity – magic board practice. If time allows, have the students break these out and spend five minutes or so working a few problems locating information from the periodic table and sharpening that skill. I have them pick these up at the start of class as they come in, so they don't get to clean up and get ready to go until they have shown me their practice. Make sure they know where to check their work (mine are on the back of the magic board binder, but you can post them wherever you like). When they are done, they should wipe down, return to the front table, etc. so they are ready for the next class.

Periodic Table Boxes Practice

	Symbol	Protons	Electrons	Mass#	Neutrons
Hydrogen					
Magnesium					
Boron					
Silicon					
Nitrogen					
Oxygen					
Iodine					
Potassium					
Barium					
Aluminium					
Tin					
Phosphorus					

© Lavish Publishing, LLC

Periodic Table Boxes Practice ANSWERS

	Symbol	Protons	Electrons	Mass#	Neutrons
Hydrogen	H	1	1	1	0
Magnesium	Mg	12	12	24	12
Boron	B	5	5	11	6
Silicon	Si	14	14	28	14
Nitrogen	N	7	7	14	7
Oxygen	O	8	8	16	8
Iodine	I	53	53	127	74
Potassium	K	19	19	39	20
Barium	Ba	56	56	137	81
Aluminium	Al	13	13	27	14
Tin	Sn	50	50	119	69
Phosphorus	P	15	15	31	16

© Lavish Publishing, LLC

Topic: Nuclear Symbols
Day: 4
Unit: Atomic Structure

Learning Target:
I can recognize and use the parts of a nuclear symbol to find the number of protons, neutrons, and electrons under a variety of sample situations.

Student Goals:
Students need to understand the difference between values found on a nuclear symbol (a specific sample) and the periodic table boxes (average values) – again this is foundational to being able to work with samples and perform calculations.

Agenda:
A – Warmup: Periodic Table Box Quiz
L – Nuclear Symbols
L – Foldable: Rutherford Model
A – Magic Board Practice

Student Materials:
Journals for notes
Foldables for Reference / Corrections
Slips of paper for quiz
Magic boards with pens and erasers

Props:
Either power point or white board to draw (I put it up as we go)

Actions and Rationale:
Warm-up – periodic table box quiz. They do the quiz while I take attendance, then I take them up for practice points before I go over them. After that, I can transition into the lesson today, which is basically two parts, so do them in whichever order you want – both parts together will be twenty to twenty-five minutes of lecture and that leaves a few minutes for magic board practice at the end.

Lecture – nuclear symbols. Up front, I start by explaining that the boxes we did yesterday are average values from samples taken on a certain date or during a certain year. Have your students pull out their reference periodic table and look for the date or year those samples were taken. All periodic tables will tell you when the sampling was done, and we need to keep in mind that those were averages, which is why we call it a reference – it's in the past.

For nuclear symbols, those values are taken from a sample we are working with right now. It is a specific sample we might have sitting in a beaker in front of us, so those values will **not** match their periodic table on those 'not perfect' values. As you see below from my recent board, these are the things we need to learn to use from our nuclear symbols:

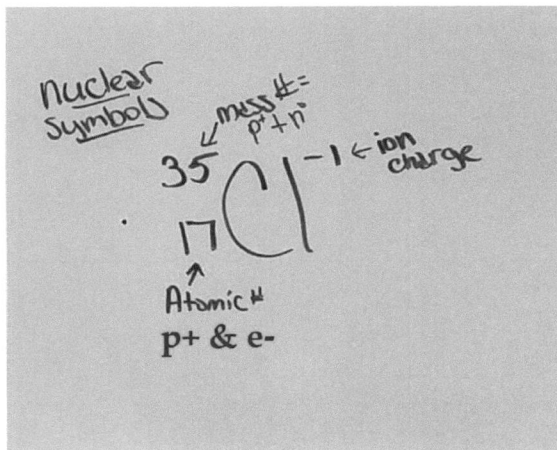

Nuclear Symbols

1. Nuclear symbols give information about a **specific sample or isotope** in a given sample
2. You need to take your information from the **nuclear symbol** - don't use the periodic table!
3. **Atomic Number** is protons & starting electrons
4. **Mass #** is protons PLUS neutrons (whole number)
5. **Ion Charge** tells the story - giver or taker and how many to add / subtract from the starting amount of e-

Just like we did yesterday, go through piece by piece and have the students give you feedback on a variety of samples you have prepared ahead of time. If you have your students pick up their magic boards as they came in, you can go through a few rows together and work out how to get each of the items for your examples. If you would rather hand these out as a worksheet, you could do that as well. And finally, whiteboards (or the blank back of the magic boards) could be used to have them provide responses so you can see if they are getting it.

You can also have them work as partners, one working a row and explaining to their partner as they go. If they get stuck, their partner can give help, then take over on the next row while the first partner listens and helps. This is a great activity, but you have to monitor, or students tend to want to take over and one do all the work while the other does nothing (hogs and logs I call them). Everyone needs to be trying.

Once you have the nuclear symbol down, you can transition to Rutherford and his gold foil experiment. Again, the portion to present for that is located on day 2 with the atomic model foldable day. Or, you can present a different model if you like. Once you are done, you are ready to have them practice for the rest of class.

Activity – magic board practice for nuclear symbols. When you are finished with the lecture, you can have the students do a few rows on their own to show you as an exit ticket. As usual, they need a way to check their work, then clean up and get ready to leave, leaving the materials ready for the next class.

Nuclear Symbol Practice

Symbol	Atomic Number Protons	Mass# - protons Neutrons	protons + neutrons Mass #	the story for e- Charge	base with story +/- Electrons
$^{1}_{1}\text{H}^{+}$					
$^{15}_{7}\text{N}^{-3}$					
$^{41}\text{Ca}^{+2}$	20				
$_{16}\text{S}^{-2}$			32		
Al	13	14		+3	
Br			80	-1	36
$^{23}_{11}\text{Na}$				+1	
Sr	38		89		36
$^{66}_{30}\text{Zn}^{+2}$					
$^{34}_{17}\text{Cl}^{-1}$					

© Lavish Publishing, LLC

Nuclear Symbol Practice ANSWERS

Symbol	Protons	Neutrons	Mass #	Charge	Electrons
	Atomic Number	Mass# - protons	protons + neutrons	the story for e-	base with story + / -
$^{1}_{1}H^{+}$	1	0	1	+1	0
$^{15}_{7}N^{-3}$	7	8	15	-3	10
$^{41}_{20}Ca^{+2}$	20	21	41	+2	18
$^{32}_{16}S^{-2}$	16	16	32	-2	18
$^{27}_{13}Al^{+3}$	13	14	27	+3	10
$^{80}_{35}Br^{-1}$	35	45	80	-1	36
$^{23}_{11}Na^{+1}$	11	12	23	+1	10
$^{89}_{38}Sr^{+2}$	38	51	89	+2	36
$^{66}_{30}Zn^{+2}$	30	36	66	+2	28
$^{34}_{17}Cl^{-1}$	17	17	34	-1	18

© Lavish Publishing, LLC

Learning Target:
I can read and annotate a science text, then answer questions about atomic structure with a regional perspective.

Student Goals:
Students gain perspective on how element availability affects regional development.

Agenda:
A – Warmup: Nuclear Symbol Quiz
A – Article Lesson

Student Materials:
Slips of paper for the quiz
Paper or Google access for questions
Copies of articles for students

Props:
Example, template or anchor chart of reading strategy

Actions and Rationale:

Warmup – nuclear symbol quiz. They do the quiz while I take attendance, then I take them up for practice points before I go over them. After that, I can transition into the lesson today, which is our reading day for this unit. It can be used if a flex day if you need more time for one of the other lectures or if you have something else you need to do, but this activity is very good for students to build their reading skills across disciplines.

Activity – article reading and annotating. Students read and annotate using the strategy that they learned during the first unit. Then they answer and submit their questions / answers. I use google classroom and a google form for collecting those responses, but paper will work just as well.

Article suggestions – Our textbook has an article that I use for this, which provides the questions for me. However, since you might have to look elsewhere, there are some good articles on the web about the history of atomic structure research and discovery.

You can make up your own questions for this, or I have had the students write their own persuasive paragraph in lieu of questions – what is the historical significance of atomic structure research? What are some inventions that have come from their discoveries? Why is continuing the research important?

Note: this is usually a Friday activity for us, mainly because a large amount of absences seem to happen on Fridays. The article is something they can do at home, and they are not missing class lecture or lab time. I also like them because it frees up some time during my day for me to work on things that I need to do while the students are still productive with an independent activity that needs little monitoring.

Topic: Vegium Lab
Day: 6
Unit: Atomic Structure

Learning Target:
I can use a manipulative lab to explore the concept and calculation of average atomic mass.

Student Goals:
Students need to know how to calculate average atomic mass. This lab is a great approximation of that process using objects that can be handled and touched, unlike actual atoms, which are an abstract concept at this point.

Agenda:
A – Warmup: Read and Question
A – Vegium Lab
A – Post Discussion of Results

Student Materials
Journals for notes
Slips of paper for quiz
Lab Sheets for each student
Vegium samples for each group
Triple Beam Balances

Props:
Completed lab to use as an example

Actions and Rationale:
Warmup – read and question. While I take attendance, I have the students read the procedures on the lab and be ready to ask any questions they have up front.

Activity – Vegium Lab. This is a popular lab that has been around for many years. If you search online, you will find many versions of it, so the one that is here is basically my version. I find that some teachers and text writers often over complicate things, and I need something that is scalable, meaning I can adjust it according to the students sitting in my room – the more SPED and ESL in a class, the simpler we need to make it.

For years, I used the version that came with my textbook, but it was extremely difficult for even my best students to complete most years, so we ended up doing the lab together as a class. Then we had ten minutes cut off of our class periods, and it became nearly impossible to complete most of our labs in only forty-eight minutes. I scaled them back for time, so feel free to modify and scale as necessary.

Samples – One thing that can take a good deal of time comes from the samples that the students get. It takes time to separate and count, so now I give them three small cups (paper 3 oz bathroom cups), one for a small amount of each 'isotope'. I show them the bag with them all mixed when I am going over the lab and explain that when a sample of an element is taken, you get all of the isotopes mixed together, but for the sake of time, I have separated them ahead of time. Also, don't make the samples too large, maybe fifteen lima beans, about fifteen or twenty grams of each of the peas and corn. They don't have to be exactly the same, and it is actually better if they are at least a little different from group to group.

Triple Beam Balances – getting the masses can be time consuming, so the more scales you have the better it will be. It doesn't have to be one per group, but if you can it will help avoid that bottle neck.

Some students need to be reminded about subtracting the cup out. They need to weigh it empty or you can give them the 'cup mass' ahead of time, but we only want the mass of the actual sample. Make a plan for this ahead of time and share it with your students.

I also have found there are a large number of students who don't know how to use the scales, so you need to be available to help, and maybe even devise a way to ensure everyone is taking a turn in the weighing – maybe a rule that each student only gets to work the scale once during the lab. Remember: hogs and logs are common if you don't have a plan for getting everyone involved.

Average Mass – now it gets tough. Students have been averaging numbers for years, but for some reason when you put a physical sample in front of them, it gets confusing. Stay cool and just remind them, average is the units you want divided by the number of items, so we are getting gram per pea or gram per bean. These are very small so they will typically be a decimal (if you have large lima beans, they might be a gram each, but I found smaller ones so the size difference wasn't quite so great).

Natural Abundance – this is a good time to mention that average atomic mass is calculated using all of the **naturally occurring isotopes**. When they divide (part divided by whole), they are putting the part they want on top and the whole sample on bottom, which gives them a decimal. Also visit the definition of **natural abundance** – abundance of isotopes of a chemical element as naturally found on a planet. The relative atomic mass (a weighted average) of these isotopes is the atomic weight listed for the element in the periodic table. It's neat to think about how that average might change if we were on Mars or Jupiter, so at some point I like to pose that question when we have a few spare minutes to ponder and discuss.

Percentage – this is also the natural abundance, more or less. They can either multiply by 100 or they can hop their decimal over two places to get there. They are finding this, but for solving the rest of the rows, they have to use the decimal (abundance) number, not this one.

Relative Weight – this is where the number of the isotopes is applied by using the natural abundance. This means that the lower number of lima beans is worth less, even though they weigh more individually. I tell my kids it would be like using the football team to decide the average weight of students at our school. How accurate would that be? Not very, and most of them understand that. We want to make it representative of the whole population, not skewed by subgroups, like the lima beans or the football team.

Post Discussion – have the students answer the questions, then bring them together for a recap. Go over anything you think is important, but I wouldn't give them the answers unless you are only doing practice points for the lab. Otherwise, I typically start with a minimum grade of fifty for completing the data table (right or wrong) and then ten points for each question they answer well as a daily grade.

If you have time, you can even do a class average on those weighted averages to get a final number per class period – then compare them across the day. Obviously, they aren't going to be exactly the same because of error, but they should be really close as a class, even if the individual groups could have quite a bit in variation depending on how close the sample cups are. If you want lots of variation, vary the amounts you are putting in the cups more, which is good for the students overall.

Finally, I tell my students that they won't have to do all that work to get average atomic mass and I will show them a shortcut tomorrow. However, it is good that they have now seen how it is done, and it works basically the same when they create the periodic table, only the measuring and counting are done a bit differently. The rest works the same and is good to know.

Vegium Lab

Our purpose today is to see how average atomic mass is calculated using a sample of vegetables – corn, split pea and lima bean. Each represents a different isotope of our new element called Vegium.

What makes isotopes of an element different?

_____

Procedure – carefully enter your data into the table. If your sample is mixed, you must separate your isotopes before you can begin.

Row 1 – mass of the isotope. Place them on the scale. If you are using a weigh boat, be sure to subtract it from your sample. The units here is grams of item (pea, corn, etc.). Add across to obtain the total.

Row 2 – number of isotopes. Count each of your isotopes, but don't lose any! Any pieces or small parts still count as one since they were weighed in the sample. Add across to obtain the total.

Row 3 – average mass. Obtained by dividing the mass of the sample by the number that is in it. The units on this number is grams per item (pea, corn, etc.). Duplicate the process to obtain the total – total mass / total parts.

Row 4 –natural abundance. Obtained by dividing the number of isotopes (part) by the number in the total sample (whole). This part / whole will give you a decimal, which is the portion that each represents out of the whole population. When you add for the total, you should get a number very close to 1.

Row 5 – percentage. Obtained by converting the abundance, which is a decimal, into a percentage, either by multiplying by 100 or moving the decimal two places to the left. When you add for the total, you should get a number very close to 100%.
Note: the formula for percentage is always **part/whole X 100**.

Row 6 – relative weight. This is commonly called a weighted average, which allows for the abundance to be considered. Obtained by multiplying the average mass x the abundance. Add across to obtain the total.

Clean up by placing your sample back in the container(s) exactly as you received it.

© Lavish Publishing, LLC

Data Table – Atomic Mass of Vegium

		Peas	Corn	Beans	Total
1.	Mass of Isotope				
2.	Number of Isotope				
3.	Average mass of Isotope				
4.	Abundance of Isotope				
5.	Percent of Isotope				
6.	Reltative Weight				

Analysis – answer each question fully using complete sentences.

1. What's the difference between 'measured' and 'calculated'?

2. Which values / rows were measured or calculated?

3. Compare the total values on row 3 and row 6. Are they the same? Why or why not?

4. Why can't atomic mass be calculated the way the total for row 3 is calculated?

5. Compare your values to those of another group in your class. What do you notice about them? Why do you think that this happened?

6. If you were to do the lab again, what would happen if you used a larger sample – would you get the same average at the end? Which would be more accurate? Explain. (Remember: accurate means closer to the actual value)

© Lavish Publishing, LLC

Topic: Calculating Average Mass
Day: 7
Unit: Atomic Structure

Learning Target:
I can calculate the average atomic mass of a given sample containing two or more isotopes.

Student Goals:
Students will need to be able to gather the necessary information from one of two sources – hyphen notation isotopes or nuclear symbols. After that, they work exactly the same, and are only about half the work of the lab, which will make it much more manageable.

Agenda:
A – Hyphen Notation Research
L – Calculating Average Atomic Mass
A – Magic Board Practice

Student Materials:
Journals for notes
Magic boards, markers and erasers for practice

Props:
Example problems with answers

Actions and Rationale:
Warmup – hyphen notation research. This doesn't need to be long or difficult, just a small paragraph or video that explains what hyphen notation is. Basically, it's the name of the element with the mass number on it, as in Carbon-12 or Carbon-14. Students can read or watch while you take attendance, then use this as a jumping off point to begin the lesson. If you want to make it a little more in depth, put a list of nuclear symbols on the board (maybe five or so) and have the students write the nuclear notation form for each. Check them together and go from there.

Lecture – calculating average atomic mass. As promised, this will be much shorter since we don't have to get the masses and abundance for ourselves. What we want to do is take the masses we are given and apply the abundance (percentage as a decimal, so that conversion we did yesterday will be necessary) to each isotope. Then add up those parts for the total. I always use the same format when making these and have an anchor chart (template) that I hang on the board to help the guide the students.

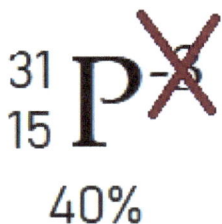

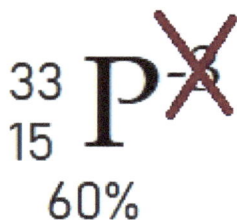

$^{31}_{15}\text{P}$~~$^{-3}$~~ $^{33}_{15}\text{P}$~~$^{-3}$~~   we don't care about the charge for this

40%   60%

mass   X   abundance   =   isotope portion

31 x .40 = 12.4

33 x .60 = 19.8

Do each line and add them all

32.2 amu

Ave Atomic Mass

I usually do a few using the symbol, then one from hyphen notation to show that they all work the same. Then we can move on to the magic board practice, which today is focused on nuclear symbols. Tomorrow's activity will be the same structure, but with hyphen notation samples.

Magic Board Practice – these can be used as guided practice as a class, a game for a team of two or four, or individual practice. Be sure to post the answers so students can check how they are doing. If you decide to run them off as a worksheet, you can still post the 'answer' and grade their work, which is how I often do it when I want more than practice points. The posted answer will guide them if they are doing it correctly either way, and the real practice will happen tomorrow when we do the Frootloop Isotopes.

Calculating Average Atomic Mass Practice

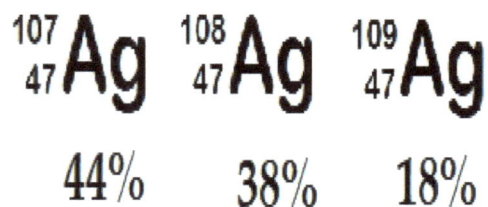

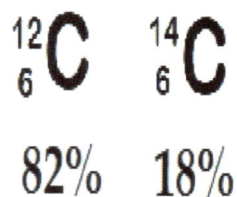

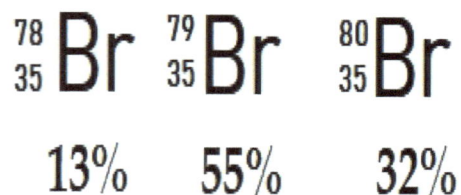

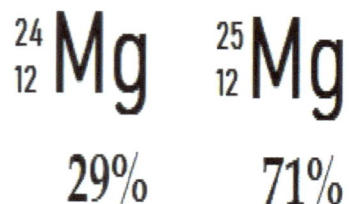

$^{107}_{47}\text{Ag}$ $^{108}_{47}\text{Ag}$ $^{109}_{47}\text{Ag}$

44%   38%   18%

$^{12}_{6}\text{C}$ $^{14}_{6}\text{C}$

82%   18%

$^{78}_{35}\text{Br}$ $^{79}_{35}\text{Br}$ $^{80}_{35}\text{Br}$

13%   55%   32%

$^{24}_{12}\text{Mg}$ $^{25}_{12}\text{Mg}$

29%   71%

© Lavish Publishing, LLC

Calculating Average Atomic Mass Practice ANSWERS

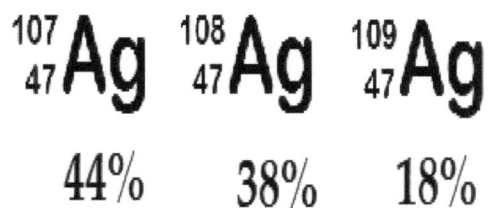

44%   38%   18%

107.74 amu

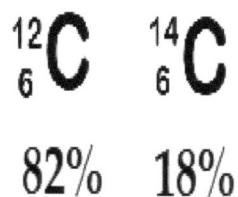

82%   18%

12.36 amu

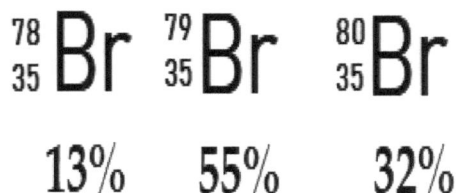

13%   55%   32%

79.19 amu

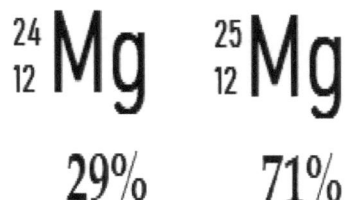

29%   71%

24.71 amu

© Lavish Publishing, LLC

Topic: Frootloop Isotopes
Day: 8
Unit: Atomic Structure

Learning Target:

I can build a basic Bohr model of my isotope to use with my group to practice calculating average atomic mass of our sample.

Student Goals:

Students need to understand the basics of atomic structures, including how to interpret hyphen notation for protons, neutrons, and electrons and relate that to calculating average atomic mass with a variety of isotopes. This activity will also solidify their understanding that isotopes are all the same element, but they have different numbers of neutrons and different masses.

Agenda:

A – Warmup: Average Atomic Mass Quiz
A – Build an Isotope
A – Calculate Average Atomic Mass

Student Materials:

Slips of paper for the quiz
White calculation sheets
Color Isotope sheets
Bottles of school glue

Props:

Finished Isotope Example (if you want one)

Actions and Rationale:

Warmup – average atomic mass quiz. This is a great checkpoint to see how they are doing with the concept, as we will be doing it all over again today with a new look or twist. They do the quiz while I take attendance, then I take them up for practice points before I go over them. On this one, I am looking for them to follow the form, not just get a number answer. After that, I transition to the activity, which is best done as a lead and follow for pause points of conversation.

Activity – Frootloop Isotopes. This is an activity I created a few years ago and have revised a little after that first try. You will have to do a bit of work for the set-up, but if you save your file afterwards, running them for subsequent years is very easy. You will also want to build one group for yourself, work the problems and get a good feel for how it works. That way you can make some tweaks to fit your students and know firsthand what the outcome should be. Also, I have it set so all the groups have four students – if you have to go with a group of three, that set of problems will have to be adjusted, such as combining seat 3 and 4 into one percentage, or having seat 4 be unbuilt, but still using that number in calculating, especially if it's a student that is absent, who will need to build and make this up later.

Preset before hand – I have two templates in here for you, one that is blank for running copies and one that is all filled in to see how it should look when completed. For the setup, you will want to make enough white copies of the blank template to make a complete class set. If you want to reuse them again next year, make one extra group beyond what you need now. The list of possible isotope groups is below, so make one set for each team. Put your hyphen notation at the top for all four for each group. Once that is done, you want to run each set or group on a different color if possible. For example, all four oxygen isotopes are red, nitrogen is blue, etc. That way each group is a different color from other groups, but they are matching each other.

Once that is set up, you will want to address your cereal. If you are in a school that they can eat them without issue, I like to just have them pull out a cup from the bag or even a bowl in the center of their group. I buy the big generic kind for this; they do their own sorting and eat their left-overs after we are all finished.

However, some schools don't allow 'food labs' so you have to plan ahead how to prevent them from eating in your class. I did this by pre-sorting the cereal into the colors, putting them into zip-lock bags for storage, and then pouring each group out two colors to work with. To prevent the eating, I told them that if the loops fell while I was sorting, I picked them up and put them in any way… and of course I have handled all of them, so bleh.

Building the isotope during class time is simple:

1. Start by having the students all **group into fours** so they are facing each other and can see each other's boards. Each group should have the same element, so I pass them out while they are transitioning into their groups (be sure to SAY the word transition when they are moving, it will come in handy later). Ask, "What is different about each of your papers?" They should notice that they all have the same element, but the number is different. "That means you all have a different **isotope** of the same element. They all have different mass numbers." This is exactly how a sample is collected for the periodic table – naturally occurring isotopes are gathered and used for the calculating. I should point out here, some of these are NOT naturally occurring, so that is something the students could put on their final answer on their work sheet.

2. Have the students fill out the **bottom of their form**, counting how many protons, neutrons, and electrons they are going to need. They will need a periodic table for this, but I have a

giant one on my wall that we use for this. You could also display one on the doc-cam or power point. When they finish, reflect, "What do you notice about everyone's numbers?" Everyone should have the same number of protons and electrons, but their neutrons should be different. Have them look at each other's boards and make sure they have subtracted correctly to get that number. "You all have a different isotope, with a different mass number, and a different number of **neutrons**." Also have the students put their name on the back of their page now, while they still can.

3. It's time to **mark our electrons**. They are going to be marking their electrons with a pen or marker. It seems to work better if you do this together. They need to put the proper number of electrons, but only put **TWO** on the first ring. After that, they put **up to** eight on each ring (which is not exactly per Bohr model, but it works for other things later), with the correct **valence number** on the valence shell. Unless you have much larger classes and have to move further down the periodic table, their electrons should fit on the three shells provided in this way. If their third shell is empty, that is good – electrons always scoot in as close to the nucleus as they can get, again called **ground state**. Share as much of this as you want while they are working, since these are all concepts you will use over and over this year.

4. Once we have the electrons marked, we can **count out cereal** for our protons and neutrons. If you are letting the students eat their leftover loops, tell them to wait until they are all done, in case they need them. Also, since they are sitting in a group, they need to share the loops to make them all right. They need the proper number for protons and neutrons, which should be **different colors**. If they glue one over the proper word at the bottom, that can act as their color key, so they will need one extra for that.

5. **Glue the model**. This part needs lots of glue if it is going to stick well, and you will need a place in your room to lay them to dry once they are done. The loops need to be pushed into the center as tightly as they can and still lay flat (Rutherford found that dense nucleus). Have the students count one more time to be sure they have the right numbers. Then ask, "What do you notice about the nucleus of your isotopes?" They should probably see that they are different sizes as they look around the table at each one.

Activity – practice average mass calculation. I always have them do this part with a **pencil** – if they make mistakes, they need to fix them until it is all right, which is a procedure we will use on calculations and worksheets for the year. If I am taking them up, they have to be right – **the work** has to be right, and this is a good time to establish that routine.

Students will need to fill in the top part first or as they build their model. Once everyone is built and ready, the group needs to work the three sets of practice problems. You will not be able to make an answer key since the seats will be flexible and change by class, but you will be able to see if they are doing it correctly. My groups have designated seats one through four all the time, so we just point out who is what seat and that is their line each time the calculation is made. Again, if you have a missing person, you can always have that page sitting in the empty desk to go in for calculating. If you run out of time from here on, the students can pick up here and finish tomorrow, or work them at home and turn them in tomorrow.

Note: Question 5 is easier because it guides them for every step of the process. For questions 6 and 7, they have to provide the 'work' and every year I have students who leave steps out or just copy their neighbor's answer. They need to practice, and I don't accept their paper until every item has been filled in correctly. As I tell my students, you can see if your average is plausible – it must fall between the lowest and highest mass number. If it is larger or smaller than the range, it is obviously wrong. I glance and if I see missing work or outside ranges, I hand it back to be redone, so we don't use our turn-in basket for this – another good routine to get used to for the year.

This will require a good deal of diligence on your part to look at their page quickly and hand it back if it isn't right, but it will be well worth it if they are working the problems out for themselves. You also have to watch out for those hogs and logs, who will let one person do all the work and everyone else just copies it, but I keep reminding them that everyone will need to know how to do it for themselves and this is the time to practice. I typically only do practice points for this page, so there really is no benefit to cheating.

Possible Isotope list:
Carbon-12, Carbon-13, Carbon-14, Carbon-15

Nitrogen-12, Nitrogen-14, Nitrogen-15, Nitrogen-16

Oxygen-16, Oxygen-17, Oxygen-18, Oxygen-19

Sodium-22, Sodium-23, Sodium-24 Sodium-25

Magnisium-24, Magnisium-25, Magnisium-26, Magnisium-27

Aluminium-25, Aluminium-26, Aluminium-27, Aluminium-28

Silicon-27, Silicon-28, Silicon-29, Silicon-30

Sulfur-32, Sulfur-33, Sulfur-34, Sulfur-36

Froot Loop Isotopes

Build your Froot Loop Model for _____.

1. The mass # of my isotope is _____.

2. My isotope has: _____ protons _____ electrons _____ neutrons.

3. Collect the correct number of froot loops for p^+ and n^o in your sample and glue them into the nucleus. Add one extra to the legend at the bottom.

4. Use a marker to draw the correct number of e^- for each of the orbitals. Mark the color on the legend.

Calculations - Use the following to calculate the average atomic mass for your group for 3 different samples.

5. Seat 1 is 35% of the sample, so the mass # of _____ x .35 = _____

  Seat 2 is 30% of the sample, so the mass # of _____ x .30 = _____

  Seat 3 is 10% of the sample, so the mass # of _____ x .10 = _____

  Seat 4 is 25% of the sample, so the mass # of _____ x .25 = _____

  Total average atomic mass is (add up the portions) _____

6. Seat 1 is 10%, so _____

  Seat 2 is 25%, so _____

  Seat 3 is 20%, so _____

  Seat 4 is 45%, so _____

  The average atomic mass is _____

7. Seat 1 is 17%, so _____

  Seat 2 is 6%, so _____

  Seat 3 is 57%, so _____

  Seat 4 is 20%, so _____

  The average atomic mass is _____

Conclusions

8. What happened to the average atomic mass of each sample?

9. Do you think they would ever be exactly the same? Why or why not?

10. Do a google search for your isotope and tell something interesting about it or what it is used for.

©Lavish Publishing, LLC

Oxygen — 15

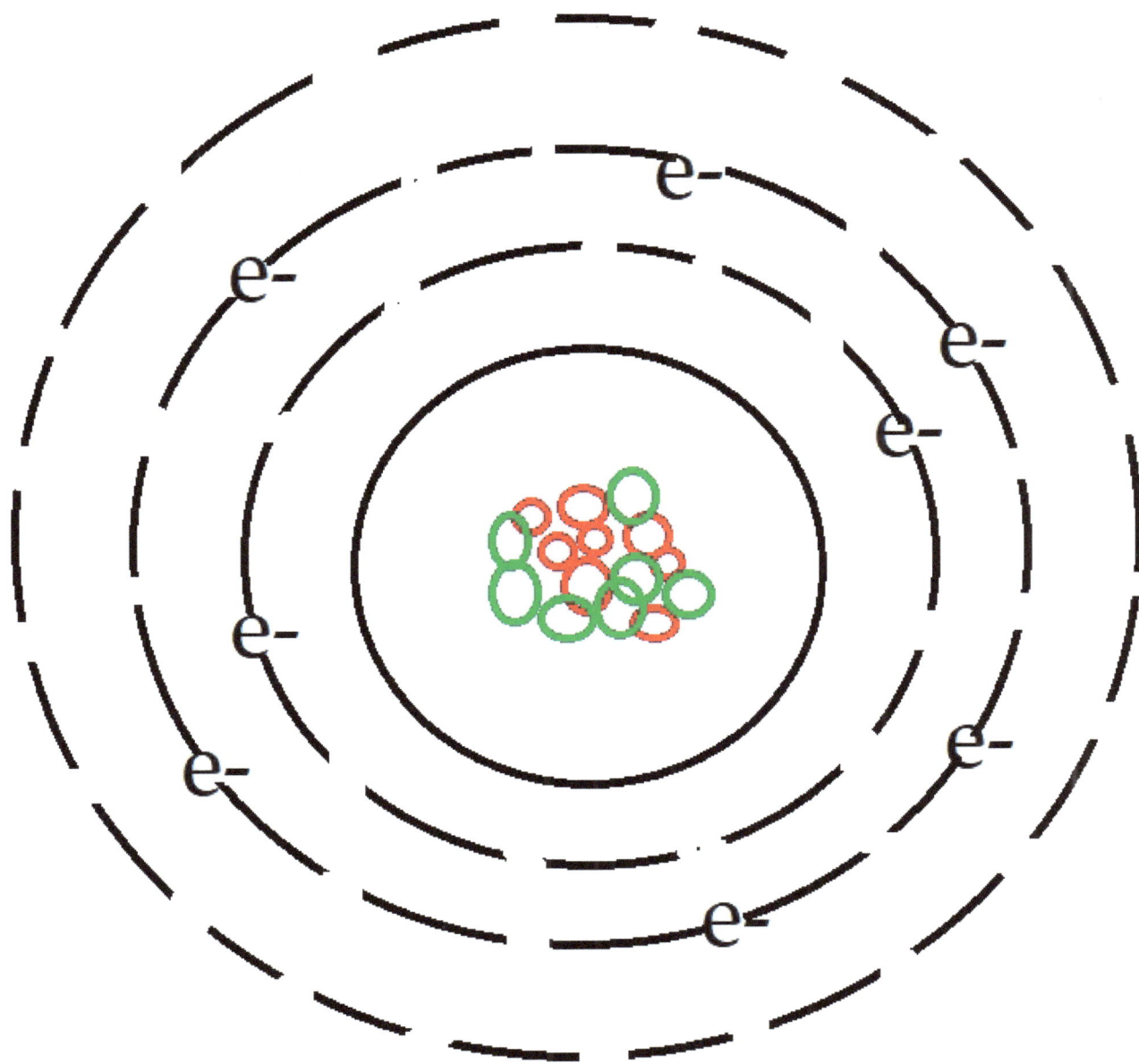

Protons = 8

Neutrons = 7

Electrons = 8

©Lavish Publishing, LLC

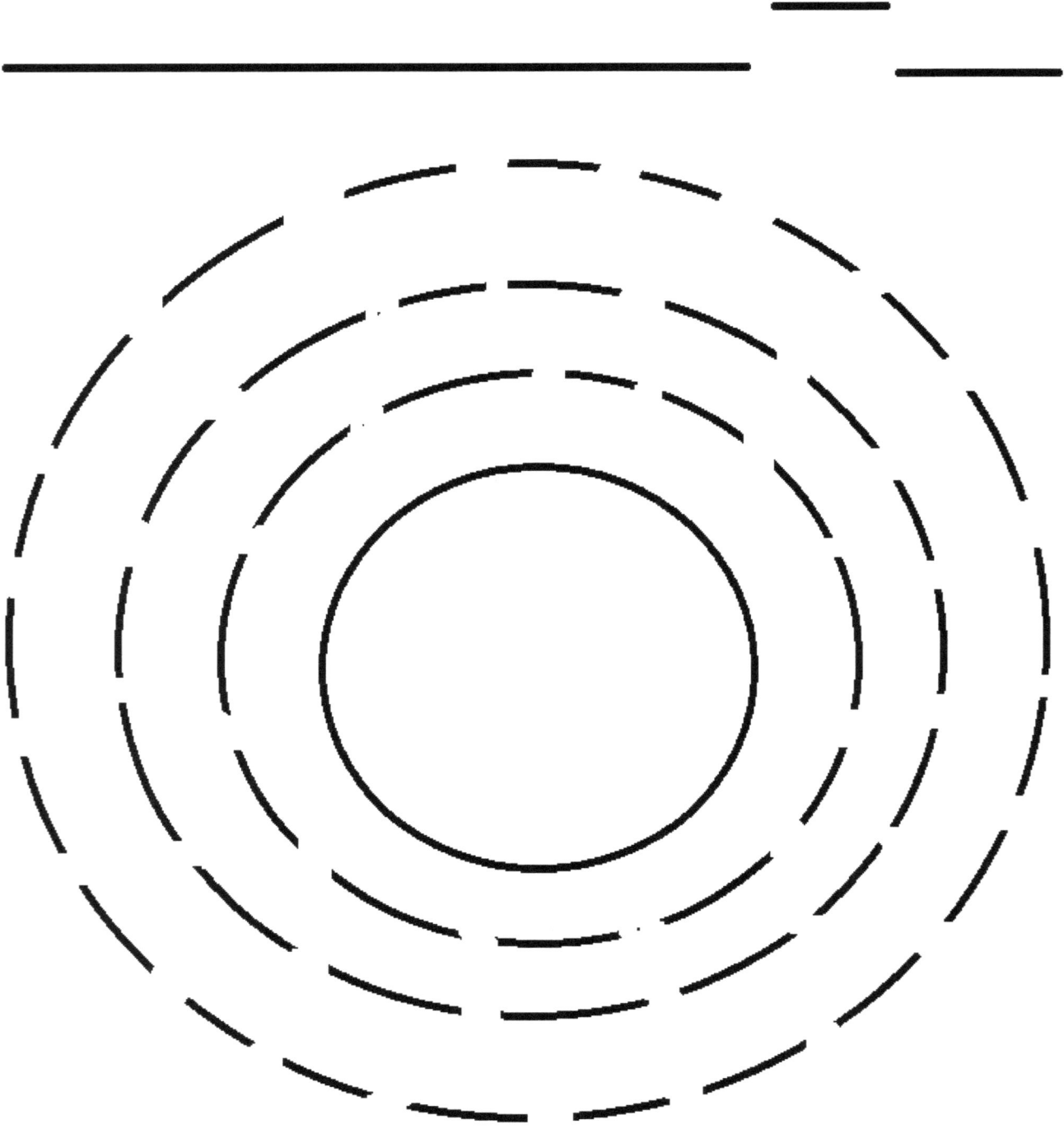

_____ _____
_____

Protons = _____   Neutrons = _____

Electrons = _____

©Lavish Publishing, LLC

Topic: Atomic Structure Review
Day: 9
Unit: Atomic Structure

Learning Target:
I can review and practice concepts that will be covered on the Properties of Matter Exam and finish any work I have missed this unit.

Student Goals:
Can name Dalton's Postulates and determine 'correctness' of each.
Can explain Thomson's experiment and contributions.
Can explain Rutherford's experiment and contributions.
Can explain Bohr's nuclear atom and contributions.
Can explain Schrodinger's experiment and quantum mechanic model.
Can explain the difference between isotopes of the same atom.
Can use nuclear symbols and calculate average atomic mass.

Agenda:
A – Magic Board Practice
A – Review Games
A – Review Sheet
A – Hand in missing work

Student Materials:
Journals and foldables for reviewing notes
Review Sheet copies for students to complete

Actions and Rationale:
Warmup – students look over the review sheet while you take attendance so they can ask questions if they need to.

Activities – magic boards. Obviously, these are some of the things we have done during the unit. Students can either make-up for when they were absent or revisit material to solidify their understanding with the review sheet to guide them. If you want to grade the review, I suggest taking it up the day of the exam so they can look over it until then.

Activity – Give Me 20 review game. This is a good unit for doing a game review, so I wanted to tell you about one of my favorites that my students have enjoyed. Typically, things that you do one or two times during the year can be quite fun, but more than that they lose a bit of their luster, so try this out if you think it fits your material and students to give it a try.

Beforehand, you will need to make your list of questions, or you can use the review sheet. For the questions, I often base them on what is going to be on the test – not the same questions, but similar. This gives me a chance to toss out any important vocab words that they are going to see, show any graphics I want them to be familiar with, and give any tidbits they might need.

This is a team game, so I have the students grouped into fours with one whiteboard that they share and take turns writing their team's response. I have my teams preplanned all year according to the seating chart, so that part is easy – they just transition (say the word transition) into their teams and we are set. My teams all have a color name, so I put the list of colors on the board for keeping score while they get grouped.

To run the game, I ask a question and they get to confer with one person writing the answer, but they need to hide it until we are ready to show – and they need to talk quietly so other teams don't steal their answers. I have a rule about asking the questions – usually I only say it once or twice, so they have to listen, then I say go and they work it out together.

When I call for the boards up, they either get a point or they don't – no writing after I call boards up (that's cheating). For the next question, we rotate the writer so they are taking turns, everyone is participating, etc. If they are noisy and we can't finish, they just don't get all of the questions, so that is incentive for them to quiet down and listen to the calling quickly between questions.

You can give out a prize to the highest score at the end if you want or give the winning team a couple of points on the test. Whatever makes you happy.

Atomic Structure Review

1. J.J. Thomas – What experiment did he perform? What did he discover?

2. What instrument can be used to observe individual atoms?

3. Which theory introduced the idea of chemical formulas?

4. What is the relative charge and mass of the three main subatomic particles?

Subatomic particle	mass	charge
proton		
neutron		
electron		

5. List Dalton's postulates – which are true and which are false. Explain why.

6. How are protons and electrons related to each other if an atom is neutral (meaning NOT an ion)?

7. Ernest Rutherford – What experiment did he perform? What did he discover?

8. How did Democritus describe atoms?

9. Find all of these using both of these sources or samples:

$^{35}_{17}Cl^{-1}$

Atomic number
Mass number
Average Atomic mass
Number of protons
Number of electrons
Number of Neutrons
Isotope Name

| 6 |
| C |
| Carbon |
| 12.01 |

©Lavish Publishing, LLC

Topic: Atomic Structure Exam
Day: 10
Unit: Atomic Structure

Learning Target:
I can demonstrate my understanding of the Properties of Matter on a written exam.

Student Goals:
Can name Dalton's Postulates and determine 'correctness' of each.
Can explain Thomson's experiment and contributions.
Can explain Rutherford's experiment and contributions.
Can explain Bohr's nuclear atom and contributions.
Can explain Schrodinger's experiment and quantum mechanic model.
Can explain the difference between isotopes of the same atom.
Can use nuclear symbols and calculate average atomic mass.

Agenda:
A – Red button day
A – Properties of Matter Exam

Student Materials:
Calculators
Scantrons or answer documents
Exam Copies

Actions and Rationale:
Red Button Day – classrooms run on routines, and this is one of my most hard fast. Basically, I have a red, yellow and green button up on my board next to the date, which is almost always on yellow. Yellow means that the students can be on their phones for APPROVED activities.

On red button day, ALL of their personal stuff goes in a designated location. I have a couple of folding tables at the front of my room that are set to about a foot tall, so they are short. Underneath, I have rugs because many kids don't like putting their bags on the floor underneath. Their bags, books, purses, etc. go on or under those tables and we do not move further until everyone has complied.

Their phones go either in their bag or on a charger, which I have two stations in my room – one that is a power strip where they use their charger, and a second set up with my personal chargers that I bought for the class to use – iPhone and android.

After we are seated and ready, I give them my testing rules –starting with the phones. We do the "pocket pat" so they can check to make sure their phones are put away and not accidently back in their pocket, and I tell them what is going to happen if they don't. I like to have them stay seated and working until everyone is finished (I do allow them to slip out to the bathroom QUIETLY). I pick up all the exams at the end, and they are not allowed to get ANY of their stuff until I have all the exams in hand, and I dismiss them to get up and prepare to leave.

Phone penalty – if I see them with a phone – using it or not – until they have been dismissed, they get a 1 on the exam (a signal that they have broken testing rules). On the first offense, I let them take the exam again during tutorials the following week and contact their parents to let them know what happened and make the offer of a retake. After that, they get a 1 and it stands. This isn't a game, and they need to learn the consequences of breaking test security.

Yes, I am very strict on this. I am not naive, and I know that kids use their phones to cheat in an unlimited number of ways when given the chance. For this one small piece of my class, I want them to show me what they know on their own, and I don't want them sharing copies of my test in any way. If you are strict and diligent, the number of problems will be far less in the long run. Be clear and up front with your expectations and stick to your rules. The first few exams they may whine and complain, but it should die down. Anyone who is over the top or causes disruption of the test over it gets a phone call home so I can chat with their parents about why their student can't follow the rules like everyone else.

For the **answer documents**, we use scantrons, which makes grading super easy, but if you don't have access, you can always have them fill out their answers on a strip of paper, which will make them easier to grade. I also like to provide each student a copy of the exam, which I always print and shrink to fit on a single page. I like to do this so they can practice writing on the test and using test taking strategies as they work. I also keep them locked up before the test, and after they are used for the rest of the year in case we need to go back and look at someone's test, so I do ask them to put their names on them.

A note about **calculators**. I do not let the students use their phones as a calculator in my class. I make them use one from the set that I provide – all the same. They practice during class time before hand, and this is the only one they get on the test. This is a good habit and I suggest it to everyone whenever possible.

Dalton's Postulates Quiz

1. Dalton described atoms as indivisible. What have we learned since?

2. Dalton said atoms could not be _____ or _____ (they are conserved).

3. Dalton said compounds are made of atoms combined in _____.

© Lavish Publishing, LLC

Dalton's Postulates Quiz

1. Dalton described atoms as indivisible. What have we learned since?

2. Dalton said atoms could not be _____ or _____ (they are conserved).

3. Dalton said compounds are made of atoms combined in _____.

© Lavish Publishing, LLC

Dalton's Postulates Quiz

1. Dalton described atoms as indivisible. What have we learned since?

2. Dalton said atoms could not be _____ or _____ (they are conserved).

3. Dalton said compounds are made of atoms combined in _____.

© Lavish Publishing, LLC

Dalton's Postulates Quiz

1. Dalton described atoms as indivisible. What have we learned since?

2. Dalton said atoms could not be _____ or _____ (they are conserved).

3. Dalton said compounds are made of atoms combined in _____.

© Lavish Publishing, LLC

Atomic Theories Quiz

1. Who drew the atom as a
nucleus surrounded by rings or shells?

2. Whose model was nick-named the
plum pudding' model?

3. Who used gold foil in his experiment?

© Lavish Publishing, LLC

Atomic Theories Quiz

1. Who drew the atom as a
nucleus surrounded by rings or shells?

2. Whose model was nick-named the
plum pudding' model?

3. Who used gold foil in his experiment?

© Lavish Publishing, LLC

Atomic Theories Quiz

1. Who drew the atom as a
nucleus surrounded by rings or shells?

2. Whose model was nick-named the
plum pudding' model?

3. Who used gold foil in his experiment?

© Lavish Publishing, LLC

Atomic Theories Quiz

1. Who drew the atom as a
nucleus surrounded by rings or shells?

2. Whose model was nick-named the
plum pudding' model?

3. Who used gold foil in his experiment?

© Lavish Publishing, LLC

PT Boxes Quiz

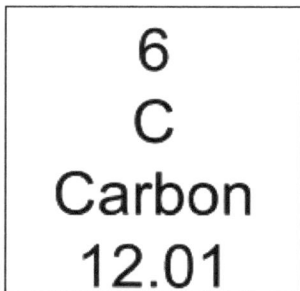

1. 6 is called the _____.

2. C is called the _____.

3. 12.01 is the _____.

© Lavish Publishing, LLC

PT Boxes Quiz

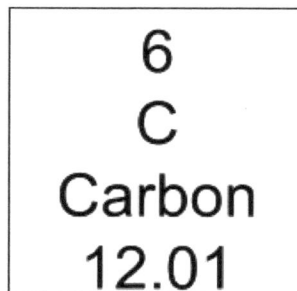

1. 6 is called the _____.

2. C is called the _____.

3. 12.01 is the _____.

© Lavish Publishing, LLC

PT Boxes Quiz

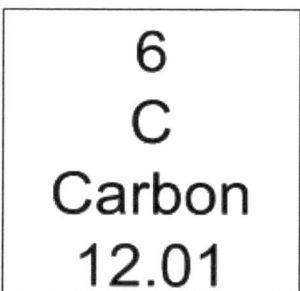

1. 6 is called the _____.

2. C is called the _____.

3. 12.01 is the _____.

© Lavish Publishing, LLC

PT Boxes Quiz

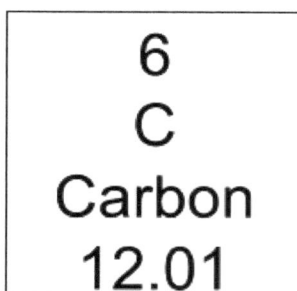

1. 6 is called the _____.

2. C is called the _____.

3. 12.01 is the _____.

© Lavish Publishing, LLC

Ave Atomic Mass Quiz

Your sample is 25% $_{6}^{12}\text{C}$

and 75% $_{6}^{14}\text{C}$

Calculate the
Ave Atomic Mass.

© Lavish Publishing, LLC

Ave Atomic Mass Quiz

Your sample is 25% $_{6}^{12}\text{C}$

and 75% $_{6}^{14}\text{C}$

Calculate the
Ave Atomic Mass.

© Lavish Publishing, LLC

Ave Atomic Mass Quiz

Your sample is 25% $_{6}^{12}\text{C}$

and 75% $_{6}^{14}\text{C}$

Calculate the
Ave Atomic Mass.

© Lavish Publishing, LLC

Ave Atomic Mass Quiz

Your sample is 25% $_{6}^{12}\text{C}$

and 75% $_{6}^{14}\text{C}$

Calculate the
Ave Atomic Mass.

© Lavish Publishing, LLC

Atomic Models Quiz

A   B   C   D

1. Which model was designed by Thomson?

2. Which model is the current standing theory?

3. Which model was created using gold foil?

© Lavish Publishing, LLC

Atomic Models Quiz

A   B   C   D

1. Which model was designed by Thomson?

2. Which model is the current standing theory?

3. Which model was created using gold foil?

© Lavish Publishing, LLC

Atomic Models Quiz

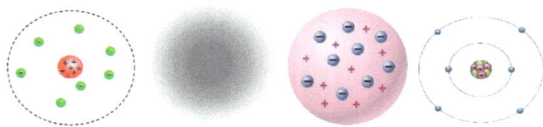

A   B   C   D

1. Which model was designed by Thomson?

2. Which model is the current standing theory?

3. Which model was created using gold foil?

© Lavish Publishing, LLC

Atomic Models Quiz

A   B   C   D

1. Which model was designed by Thomson?

2. Which model is the current standing theory?

3. Which model was created using gold foil?

© Lavish Publishing, LLC

Nuclear Symbols Quiz

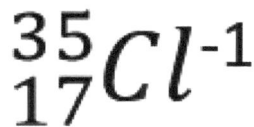

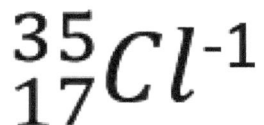

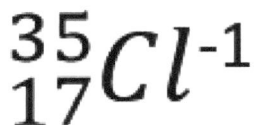

$^{35}_{17}Cl^{-1}$

1. What is the mass number?

2. How many protons?

3. How many electrons?

4. Name this isotope (use hyphen notation).

© Lavish Publishing, LLC

Nuclear Symbols Quiz

$^{35}_{17}Cl^{-1}$

1. What is the mass number?

2. How many protons?

3. How many electrons?

4. Name this isotope (use hyphen notation).

© Lavish Publishing, LLC

Nuclear Symbols Quiz

$^{35}_{17}Cl^{-1}$

1. What is the mass number?

2. How many protons?

3. How many electrons?

4. Name this isotope (use hyphen notation).

© Lavish Publishing, LLC

Nuclear Symbols Quiz

$^{35}_{17}Cl^{-1}$

1. What is the mass number?

2. How many protons?

3. How many electrons?

4. Name this isotope (use hyphen notation).

© Lavish Publishing, LLC

1		____ are negatively charged ions - atoms that take electrons.
2		____ are subatomic particles with a negative charge, no mass value, and are located outside the nucleus of the atom; responsible for chemical reactions.
3		____ attempt to explain the structure of an atom and electron configuration by using the laws of probability to predict the location of electrons.
4		____ is a planetary model in which the negatively charged electrons orbit a small, positively-charged nucleus similar to the planets orbiting the Sun (except that the orbits are not planar)
5		____ is equal to protons plus neutrons; it is the average atomic mass rounded to a whole number.
	word choices	(A) anion (B) Bohr's Atomic Model (C) electron (D) mass number (E) Schrodinger / quantum model

1		By taking the weighted percentage of all naturally occurring isotopes of an element, we can calculate the ____, which is the mass found on the periodic table.
2		____ is equal to the number of protons in an atom; used to arrange the periodic table.
3		The outer energy level of an atom is called the ____ shell and the electrons that are in it are called ____ electrons. (same word for both blanks).
4		The parts that make up an atom are called ____ - protons, neutrons and electrons.
5		____ are positively charged ions - atoms that give away electrons.
	word choices	(A) atomic number (B) average atomic mass (C) cations (D) subatomic particles (E) valence

© Lavish Publishing, LLC

1		Atoms have different numbers of neutrons in their nuclei, which are said to be _____ because their masses will vary.
2		A subatomic particle with a positive charge is called a ____, has mass of about 1, and is located in the nucleus of the atom.
3		In ____, he used a cathode ray to determine charges and locations of subatomic particles. His theory (called the 'plum pudding model) was later disproved, but did serve as a strong catalyst for the work of others.
4		In the early 1800's, ___ were published, which gave us a strong definition for nuclear theory, much of which is upheld today.
5		Atoms that give or take electrons and become ions are said to have a positive or negative ____, which causes them to attract or repel other particles that have a ____similar to a magnet (same word in both blanks).
	word choices	(A) charge (B) Dalton's Postulates (C) isotopes (D) proton (E) Thomson's Experiments

Form D **Atomic Structure Vocabulary Quiz**
© Lavish Publishing, LLC

1		The _____ tells us that all atoms want to have 8 electrons in their valence shell. Remember that all atoms can only hold 2 electrons in the very first energy level, which is the only completely stable exception to this rule, such as Hydrogen or Helium.
2		_____ are the result of gaining or losing electrons by atoms.
3		___ is equal to protons plus neutrons; it is the average atomic mass rounded to a whole number
4		By taking the weighted percentage of all naturally occurring isotopes of an element, we can calculate the ____, which is the mass found on the periodic table.
5		The letter or letters that represent an element are called the _____, which can include details about the structure or parts of that atom.
	word choices	(A) average atomic mass (B) ions (C) mass number (D) nuclear symbol (E) octet rule

Atomic Structure Exam

Use this symbol to answer questions 1 – 3.

$$^{33}_{15}P^{-3}$$

1. How many protons does this atom of phosphorous have?
A. 12   C. 18
B. 15   D. 33

2. How many electrons would this atom of sulfur have?
A. 12   C. 18
B. 15   D. 33

3. How many neutrons would this atom of sulfur have?
A. 12   C. 18
B. 15   D. 33

4. An isotope of lead has 82 protons and 115 neutrons. What is the mass number of this isotope?
A. 82   C. 197
B. 115   D. 207

5. Which of the following is true about Nitrogen-17? It has…
A. 7 neutrons   C. 17 electrons
B. 7 protons   D. 17 neutrons

6. Which of the following is used to calculate the average atomic mass of an element?
A. The most common isotopes
B. The five most abundant isotopes
C. The isotopes that are man made
D. The naturally occurring isotopes

7. What is the mass number of a sodium isotope that consist of 11 protons, 14 neutrons, and 11 electrons?
A. 11   C. 25
B. 14   D. 36

©Lavish Publishing, LLC

8. Isotopes of atoms vary by having...
A. Different masses
B. Different numbers of protons
C. Different chemical properties
D. Different numbers of electrons

9. How many electrons does Mn have?
A. 22   C. 36
B. 25   D. 54

10. Using the Periodic Table, predict which elements will have similar chemical properties or reactivity.
A. Cobalt, Calcium and Carbon
B. Nitrogen, Silicon and Gallium
C. Cesium, Barium, and Halfnium
D. Fluorine, Chlorine and Bromine

11. As the atomic number on the Periodic Table increases, the number of electrons...
A. is unchanged   C. decreases
B. alternates   D. increases

12. As the atomic number on the Periodic Table increases, the number of protons...
A. is unchanged   C. decreases
B. alternates   D. increases

13. Henry Moseley studied x-ray spectra of several elements in a row of the Periodic Table. He found that each element had one more proton than the element immediately to its left.

Which statement best describes Moseley's contribution to the modern Periodic Table?

A. Providing a basis for ordering elements in the Periodic Table by atomic number
B. Arranging elements in the Periodic Table based on their properties
C. Recognizing that elements in the Period Table have similar properties
D. Predicting the properties of missing elements in the Periodic Table

14. What is group 18 (group A8) on the Periodic Table called?
A. Alkali Metals   C. Halogens
B. Alkaline Earth Metals   D. Noble Gases

15. Which of the following showed electrons in specific, fixed orbitals?
A. Rutherford Model   C. Bohr Model
B. Planck Model   D. Quantum Model

© Lavish Publishing, LLC

16. How many electrons does Ca^{+2} have?

A. 2   C. 20

B. 18   D. 22

17. Some physical properties of elements in a group on the Periodic Table are described below.

       Good conductor of electricity
       Good conductor of thermal energy (heat)
       Atoms contain two valence electrons
       Malleable in their solid state

Which family is being described?

A. Alkaline Earth Metals   C. Halogens

B. Alkali Metals   D. Noble Gases

18. What is the charge of Group 7 (Group A7) elements?

A. +1   C. -2

B. -1   D. -3

19. Which atomic model was discovered using gold foil?

A. Thomson   C. Bohr

B. Rutherford   D. Schrodinger

20. Which family on the periodic table is the most reactive?

A. Group 1   C. Group 17

B. Group 2   D. Group 18

© Lavish Publishing, LLC

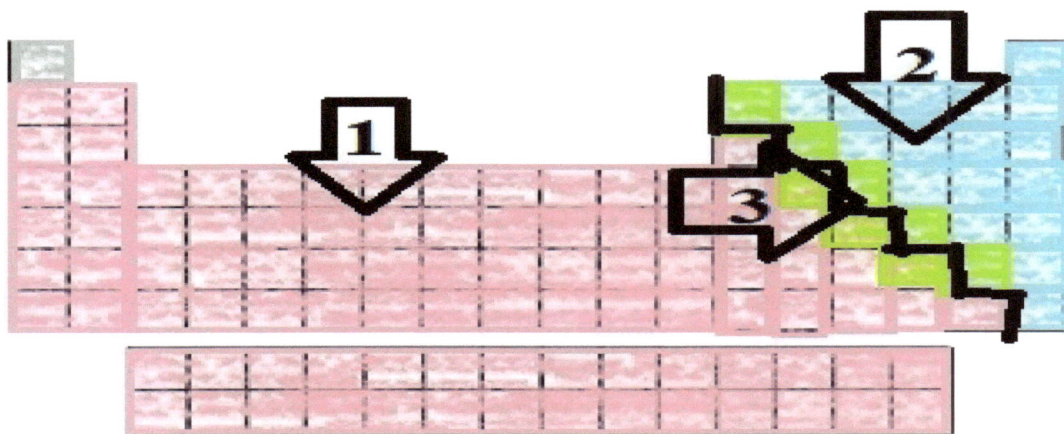

Use the periodic table above for questions 21 & 22.

21. Which section consists of the non-metals?
A. 1   B. 2   C. 3   D. none of them

22. Which section consists of the metals?
A. 1   B. 2   C. 3   D. none of them

23. Whose theory does   represent?
A. Thomson   C. Bohr
B. Rutherford   D. Schrodinger

24. Whose theory does   represent?
A. Thomson   C. Bohr
B. Rutherford   D. Schrodinger

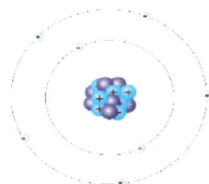

25. Whose theory does   represent?
A. Thomson   C. Bohr
B. Rutherford   D. Schrodinger

© Lavish Publishing, LLC

Dalton's Postulates Quiz

1. Dalton described atoms as indivisible. What have we learned since?
actually of protons, neutrons & electrons

2. Dalton said atoms could not be _____ or _____ (they are conserved).
created or destroyed

3. Dalton said compounds are made of atoms combined in _____.
whole number ratios

© Lavish Publishing, LLC

Atomic Theories Quiz

1. Who drew the atom as a nucleus surrounded by rings or shells?
Niels Bohr

2. Whose model was nick-named the plum pudding' model?
JJ Thomson

3. Who used gold foil in his experiment?
Ernest Rutherford

© Lavish Publishing, LLC

PT Boxes Quiz

6
C
Carbon
12.01

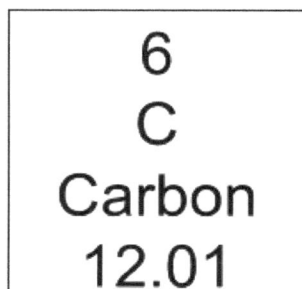

1. 6 is called the _____.
atomic number
2. C is called the _____.
symbol
3. 12.01 is the _____.
average atomic mass

© Lavish Publishing, LLC

Ave Atomic Mass Quiz

Your sample is 25% $^{12}_{6}C$ and 75% $^{14}_{6}C$

Calculate the Ave Atomic Mass.

$12 \times .25 = 3$
$14 \times .75 = 10.5$
$3 + 10.5 =$ **13.5 amu**

© Lavish Publishing, LLC

Atomic Models Quiz

A   B   C   D

1. Which model was designed by Thomson?

C

2. Which model is the current standing theory?

B

3. Which model was created using gold foil?

A

© Lavish Publishing, LLC

Nuclear Symbols Quiz

$$^{35}_{17}Cl^{-1}$$

1. What is the mass number?

35

2. How many protons?

17

3. How many electrons?

18

4. Name this isotope (use hyphen notation).

Chlorine-35

© Lavish Publishing, LLC

Atomic Structure Vocabulary Quiz Answers

	Form A		Form B		Form C		Form D
1	A Anion	1	B Average Atomic Mass	1	C Isotopes	1	E Octet Rule
2	C Electron	2	A Atomic Number	2	D Proton	2	B Ions
3	E Sc hrodinger / quantum model	3	E Valence	3	E Thomson's Experiments	3	C Mass Number
4B	B Bohr's Atomic Model	4	D Subatomic Particles	4	B Dalton's Postulates	4	A Average Atomic Mass
5	D Mass Number	5	C Cations	5	A Charge	5	D Nuclear Symbol

Atomic Structure Exam Answers

1. B	6. D	11. D	16. B	21.B
2. C	7. C	12. D	17. A	22. A
3. C	8. A	13. A	18. B	23. A
4. C	9. B	14. D	19. B	24. D
5. B	10. D	15. C	20. A	25. C

About the Author

Born and raised in West Texas, Sammie Jacobs aspired to be a teacher at an early age but did not achieve her dream until the age of 38 when she earned her composite science certification in 2008.

Answering the call of the classroom, she went to work at a local high school teaching Chemistry. Through all the years, she loved the subject and her students. Combining her dedication to both, Sam continuously searched for ways to improve her instruction and meet her students varied needs, leading her to create and construct many of her materials.

Mentoring new teachers as they came into the district and her department, Sam could see the importance of helping those new to her beloved field. First, she began to build the files that would become the foundation of this series. Later, she realized that putting those lesson plans and tools into the hands of others could mean a great deal to her colleagues and countless students, eventually even those across the country or around the world.

Always striving for more and looking for the best in herself and those around her, Sammie Jacobs is releasing this complete version of her Chemistry lessons – 17 units in all, which will be available for purchase at a nominal fee in paperback format. These files contain everything needed to plan and execute a solid foundational year of Chemistry for any teacher, be it veteran or novice.

But Sam also knows there is more that can be done, so she is also founding a teacher community group on Facebook. Open to everyone who works as a science educator, this is Sam's legacy, her dream coming true. If you are a veteran with advice to share or a novice looking for a helpful hand, come and join and let us grow together. In the end, Sam hopes many will find these tools useful and many more will be inspired to reach for their dreams, whatever they might be…

Units in the Teaching Chemistry
in a Diversified Classroom Series

Lab Safety & Equipment
Properties of Matter
Periodic Table
Atomic Structure
Electrons in Atoms
Ionic Bonding
Covalent Bonding
Names & Formulas
Equations & Reactions
Calculations
Stoichiometry
States of Matter
Gases
Solutions
Acids & Bases
Thermochemistry
Nuclear Chemistry

Follow the Teaching Chemistry Series on Facebook –
https://www.facebook.com/SammieJacobsChemistry/?

Join the Chemistry Professional Learning Community on Facebook –
https://www.facebook.com/groups/TeachingChemistryPLC/

www.ingramcontent.com/pod-product-compliance
Lightning Source LLC
Chambersburg PA
CBHW042010080426

42734CB00002B/31